THE STUFF OF LIFE

THE CHANDLER LECTURE
1935

THE STUFF OF LIFE

BY

JACOB G. LIPMAN

Agricultural Experiment Station, New Brunswick, N. J.

NEW YORK

COLUMBIA UNIVERSITY PRESS

1935

[Reprinted from Industrial and Engineering Chemistry,
Vol. 27, page 103, January, 1935]

Printed from type January, 1935

Printed by Mack Printing Co.
Easton, Pa., U. S. A.

J. G. LIPMAN

The Stuff of Life

JACOB G. LIPMAN, Agricultural Experiment Station, New Brunswick, N. J.

The Chandler Lecture for 1934 was delivered at Columbia University, New York, on December 14, 1934, by Jacob G. Lipman, Dean of Agriculture and Director of the New Jersey Agricultural Experiment Station at Rutgers University. In presenting the medal awarded at the close of the lecture, Dean H. L. McBain of Columbia spoke in part as follows in praise of the medalist:

A native of Latvia, then a part of the Russian Empire, but early emigrating with his parents to the United States and settling in the State of New Jersey, Dr. Lipman received his education at Rutgers College, an institution which he has long honored, and at Cornell University where he pursued graduate studies leading to the degrees of master of arts and doctor of philosophy in the field of agricultural chemistry. From Cornell he returned to Rutgers to teach and to pursue researches which have had far-reaching consequences of both theoretical and practical value. His researches on the utilization of nitrogen by plants, especially upon the reactions that influence the availability of nitrogen to plants, are outstanding contributions, as are his studies of the soil chemistry of sulfur, phosphorus, selenium, and other chemicals. His discovery of the nature of bacteriological action in soils has been of the highest practical significance.

These researches have given him an enviable, world-wide reputation in the important field of agricultural chemistry. It is because of his achievements in this field that he was selected to receive the award of the Chandler Medal.

He was the founder of the first journal of soil science in America, first president of the International Society of Soil Science (in 1924), president of the First International Congress of Soil Science (in 1927), representative of the Government of the United States at many international agricultural congresses and conferences, and has been honored by scientific societies and leading universities the world over.

The Charles Frederick Chandler Foundation was established in 1910, when friends of Professor Chandler presented to the trustees of Columbia University a sum of money, and stipulated that the income was to be used to provide a lecture by an eminent chemist and also a medal to be presented to this lecturer in further recognition of his achievements in the chemical field.

The previous lecturers and the titles of their lectures are as follows (all have been published in INDUSTRIAL AND ENGINEERING CHEMISTRY):

1914	L. H. Baekeland	Some Aspects of Industrial Chemistry
1916	W. F. Hillebrand	Our Analytical Chemistry and Its Future
1920	W. R. Whitney	The Littlest Things in Chemistry
1921	F. G. Hopkins	Newer Aspects of the Nutrition Problem
1922	E. F. Smith	Samuel Latham Mitchill—A Father in American Chemistry
1923	R. E. Swain	Atmospheric Pollution by Industrial Wastes
1925	E. C. Kendall	Influence of the Thyroid Gland on Oxidation in Animal Organism
1926	S. W. Parr	The Constitution of Coal—Having Special Reference to the Problems of Carbonization
1927	Moses Gomberg	Radicals in Chemistry, Past and Present
1928	J. A. Wilson	Chemistry and Leather
1929	Irving Langmuir	Electrochemical Interactions of Tungsten, Thorium, Caesium, and Oxygen
1931	James B. Conant	Equilibria and Rates of Some Organic Reactions
1933	George O. Curme, Jr.	Synthetic Organic Chemistry in Industry

LIFE is clad in many garments, all made from the same stuff. In the air, the sea, and the earth, plant and animal cells find the raw materials for weaving the web of life. The Great Weaver we know but do not understand. The loom we can build and destroy. The fabric we possess, and, in the possessing, there come to us fear and envy. The poet may think of us as "such stuff as dreams are made on;" the historian, as dwarfs and giants moving across the stage; the philosopher, as creatures of our environment and of elemental forces hidden within ourselves; the biologist, as protoplasm stretching from the beginning to the end of biological time. The chemist may accept all of these appraisals and definitions as not running counter to his own conception of living cells as foci of matter and energy or, shall we say, of energy alone.

Carlyle says in his "Sartor Resartus" (*8*):

To the eye of vulgar Logic, what is man? An omnivorous Biped that wears Breeches. To the eye of Pure Reason, what is he? A Soul. A spirit, and divine Apparition. Round his mysterious Me, there lies, under all those wool-rags, a Garment of Flesh (or of Senses), contextured in the Loom of Heaven; whereby he is revealed to his like, and dwells with them in Union and Division; and sees and fashions for himself a Universe, with azure Starry Spaces, and long Thousands of Years. Deep-hidden is he under that strange Garment; amid Sounds and Colours and Forms, as it were, swathed-in, and inextricably over-shrouded: yet it is skywoven, and worthy of a God....... That living flood pouring through these streets, of all qualities and ages, Knowest thou whence it is coming, whither it is going? *Aus der Ewigkeit, zu der Ewigkeit hin:* From Eternity, onward to Eternity! These are Apparitions: what else? Are they not souls rendered visible; in Bodies, that took shape and will lose it, melting into air? Their solid pavement is a Picture of the Sense; they walk on the bosom of Nothing, blank Time is behind them and before them.......Friend, thou seest here a living link in that Tissue of History, which inweaves all Being; watch well, or it will be past thee, and seen no more.

In the philosophy of Aristotle matter is derived from four primary elements—fire, air, earth, and water. For us, the fires of the sun are still burning; and air, water, and earth are still the source of raw materials for building the tissues of living organisms. Directly and indirectly, atmospheric gases provide oxygen, nitrogen, and carbon. Water furnishes oxygen and hydrogen. Earth is the source of many elements found in plants and animals. Some of these elements are indispensable for cell development, others are incidental in nutrition processes. Parenthetically, it may be noted that the list of elements known to be essential for plant growth is substantially longer than it was a few years ago.

The garment of life is of many patterns, but the building materials are the same. The "why" of the pattern we do not fully understand, even though we have accumulated and interpreted a vast fund of information on the chemistry of carbon compounds. We acknowledge our debt to the chemist for telling us what happens. Then we turn to the biologist and ask him, with the barest shadow of a smile, whether protoplasm has a memory. Certainly, it shuffles its building materials with astonishing versatility.

Let us consider, for a moment, the quantitative relations of the simple ingredients that enter into the building of plant and animal tissues. Our purpose will be served if we compare the average analyses, on a water-free basis, of important whole plants and of the entire bodies of steer and man.

The analytical data on plants, animals, and portions of them, have provided the composite as given in Table I. They have been culled from a wide variety of sources including those furnished by the New Jersey Experiment Station (5).

TABLE I. COMPOSITION OF ANIMAL AND VEGETABLE MATTER

	MAN %	ANIMAL (STEER) %	PLANT %	TREE %
Water	65.00	48.89	70.33	43.82
Dry matter	35.00	51.11	29.67	56.18
	DRY BASIS			
Ash (as oxides)	19.12	8.06	6.00	0.54
Organic matter (including all oxygen and carbon)	88.18	95.28	96.14	99.64
Carbohydrates	2.94	3.16	78.99	97.64
Fats	41.25	48.32	4.27	1.25
Proteins	43.99	43.80	12.88	0.75
Carbon	54.21	63.10	43.60	49.99
Oxygen	18.88	16.37	44.54	43.40
Hydrogen	8.04	7.89	6.25	6.05
Nitrogen	7.05	7.92	1.75	0.20
Ash as inorganic elements	11.82	4.72	3.86	0.36
Calcium	6.38	2.55	0.25	0.08
Magnesium	0.18	0.10	0.59	0.03
Potassium	0.81	0.14	1.08	0.08
Sodium	0.38	0.12	0.18	0.02
Iron	0.03	0.003	0.09	0.01
Aluminum	0.0001	0.0001	0.08	0.07
Sulfur	0.53	0.22	0.81	0.01
Phosphorus	3.33	1.39	0.39	0.01
Manganese	0.001	0.001	0.04	0.02
Silicon	0.005	0.005	0.29	0.01
Chlorine	0.18	0.19	0.06	0.015

Age and species of plants and animals account for rather wide differences in their water content. The percentage of fat is a significant factor in the case of animals. Attention may also be called to differences in the water content of plant and tree. Evidently, the percentage of dry matter must vary directly with that of the uncombined water.

There is a wide difference in the ash content of man and animal. The weight of the bones and their composition, as well as the proportion of fat in the body, will largely account for this difference. The ash (calculated as oxides) in the dry substance of man, animal, plant, and tree is equivalent to 19.12, 8.06, 6.00, and 0.54 per cent, respectively.

The organic portion of man and animal consists of almost equal proportions of fats and proteins. The proportion of carbohydrates is relatively small. On the other hand, the dry matter of plant and tree is high in carbohydrates and relatively low in protein and fat. The variations in the percentage of these proximate constituents will be noted later.

The major mineral constituents of the animal body are calcium and phosphorus. Roughly, there are present two parts of calcium to one of phosphorus. The other ash ingredients are present in small or minute amounts, but they are important, nevertheless. More will be said on this point later. Plant ash contains a relatively high proportion of potassium. As between plants and animals, differences will be found also in the case of calcium, magnesium, sodium, chlorine, iron, aluminum, manganese, sulfur, and silicon.

One of the striking facts about the dry substance of plants and animals is its high content of carbon. The source of this element for plants is the atmosphere. Indirectly, it is the source for animals also. Atmospheric air contains only about 0.03 per cent of carbon dioxide. Despite its extreme dilution, plants use the carbon dioxide of the atmosphere as the source of their major constituent, carbon. Dry plant substance contains rather less than 50 per cent of carbon; dry animal substance contains more than 50 per cent. The figures given by Hart and Tottingham (*15*) for the percentage composition of starch and plant stearin are as follows:

	CARBON	HYDROGEN	OXYGEN
Carbohydrate (starch)	39.98	6.71	53.31
Fat (stearin)	76.78	12.45	10.77

They also note a variation in the fat content of some common seeds, ranging from about 2 per cent in wheat, 5 per cent in maize, 20 per cent in cotton, 33.5 per cent in flax, to 50 per cent in the castor bean. The range of variation as to proteins, carbohydrates, and ash is no less significant. According to Henry and Morrison (*16*), the average of many analyses of the seed of soy beans and rice, for example, indicates a protein content of 7.6 per cent for rice and 36.5 for soy beans. The corresponding percentages of carbohydrates were 30.8 for soy beans and 76.0 for rice. As to fat, it was 1.9 per cent for rice and 17.5 for soy beans. The ash content was as low as 1.5 per cent in corn and as high as 5.3 in soy beans.

Similar analyses are given by Henry and Morrison for such

forage crops as alfalfa, timothy, and pasture grasses. We find here a range of 6.2 to 14.9 per cent of protein in timothy and alfalfa, respectively. The corresponding range is 74.8 to 71.5 per cent for carbohydrates in timothy and pasture grasses, respectively. The fat content is about the same in each case—namely, 2.5 per cent. The proportion of ash is a little less than 5 per cent in timothy and somewhat more than 8.5 in alfalfa.

If we consider these and other analytical data in somewhat diagrammatic fashion, we find that the soil contributes about 5 per cent as ash, and air and water the remaining 95 per cent of the dry substance of plants. The protein varies, let us say, from 5 to 35, carbohydrates from 30 to 75, and fat from 2 to 20 per cent. If we assume an approximate carbon content for fat, protein, and carbohydrates of 76, 52, and 44 per cent, respectively, we discover that dry plant substance contains, roughly, 40 to 50 per cent of carbon. The corresponding range in animal bodies is, let us say, 50 to 60 per cent. What are the sources of this carbon, and what, in general, are the earth's resources of the various raw materials of plant and animal organisms?

Sources of Raw Materials

Clarke (*10*) envisions for us the changes in the surface of the Plutonic rocks on the earth's crust. He recognizes three shells or layers. The first is Plutonic rock of unknown thickness, overlaid, successively, by shells of sedimentary and fragmentary rocks and by unconsolidated material composed of soils, sands, clays, and gravels. The limits of these shells are not rigid; they shade into one another and they represent surface material in the making of which air, water, and living organisms have played a part. There has been the absorption of carbon and oxygen from the atmosphere and hydration of clays and shales. There has also been the leaching of salts and their removal to the ocean. "When igneous rocks," says Clarke, "are transformed into sedimentary rocks, there is an average net gain of weight of 8 or 9 per cent, as roughly estimated from the composition of the various kinds of rock under consideration. To some extent, then, the ocean and the atmosphere are being slowly absorbed by and fixed in the solid crust of the earth, although under certain conditions this tendency is reversed, with liberation of water and of gases." Tables II and III are quoted from Clarke (11).

TABLE II. COMPOSITION OF KNOWN MATTER OF THE EARTH

Density of crust	2.5	2.7
Atmosphere, %	0.03	0.03
Ocean, %	7.08	6.58
Solid crust, %	92.89	93.39
	100.00	100.00

TABLE III. PRINCIPAL CONSTITUENTS OF THE ATMOSPHERE

	BY WEIGHT	BY VOLUME
Oxygen	23.024	20.941
Nitrogen	75.539	78.122
Argon[a]	1.437	0.937

[a] This includes krypton, xenon, helium, and neon.

The other constituents of the atmosphere include aqueous vapor, carbon dioxide, ammonia, sulfur compounds, ozone, nitrates and nitrites, dust particles, etc. Reference will be made later to the quantitative significance of oxygen, nitrogen, carbon, hydrogen, and other elements in relation to their origin.

TABLE IV. AVERAGE COMPOSITION OF KNOWN TERRESTRIAL MATTER

ELEMENT	LITHO-SPHERE (93%)[a] %	HYDRO-SPHERE (7%) %	AVERAGE[b] %
Oxygen	46.46	85.79	49.20
Silicon	27.61	...	25.67
Aluminum	8.07	...	7.50
Iron	5.06	...	4.71
Magnesium	2.07	0.14	1.93
Calcium	3.64	0.05	3.39
Sodium	2.75	1.14	2.63
Potassium	2.58	0.04	2.40
Hydrogen	0.14	10.67	0.87
Titanium	0.62	...	0.58
Carbon	0.09	0.002	0.08
Chlorine	0.05	2.07	0.19
Bromine	...	0.008	...
Fluorine	0.03	...	0.03
Phosphorus	0.12	...	0.11
Sulfur	0.06	0.09	0.06
Manganese	0.09	...	0.09
Barium	0.04	...	0.04
Strontium	0.02	...	0.02
Nitrogen	...	...	0.03
All others	0.50	...	0.47
	100.00	100.00	100.00

[a] More correctly, the 10-mile crust. [b] Including the atmosphere.

Two elements, oxygen and silicon, make up nearly 75 per cent of the known terrestrial matter (Table IV). If we add aluminum and iron to the list we obtain a total of about 87 per cent. This is increased to nearly 92.5 per cent by the addition of calcium and magnesium, and to approximately 97.5 by the further addition of sodium and potassium. In other words, all but 2.5 per cent of known terrestrial matter is made up of these eight elements. The other elements on the list include carbon which furnishes about 40 to 50 per cent of the dry matter of plants and animals. They include, also, hydrogen, nitrogen, phosphorus, sulfur, manganese, and several others without which plant and animal tissues cannot be built. It is particularly worth noting that, with the exception of oxygen, the four elements at the top of the list constitute only a minor portion of organic matter. Obviously, living organisms can use the constituents of the atmosphere and the lithosphere in their own way for the building of cell patterns, and always it is carbon that serves as the keystone of the arch.

Carbon Resources

Reference has already been made to the fact that plants contain, in general terms, 40 to 50, and animals 50 to 60 per cent of carbon in their dry substance. Enormous quantities of carbon dioxide are used daily in the plant world as a primary source of carbon to vegetation. The latter, in its turn, serves as a direct or indirect source of carbon for animals. But since plants obtain their carbon from the atmosphere, or from atmospheric gases dissolved in water, and, since the atmosphere contains only 0.03 per cent by volume of carbon dioxide in dry air, it is obvious that there must be a rapid turnover in the current supply of carbon for plants and animals. Numerous studies and calculations on the subject of carbon in its relation to living organisms are reported in widely scattered literature. A summary and interpretation of the available data will be found in a rather recent book by Vernadsky (*42*). His figures deal with the carbon content of the atmosphere and the additions to be credited to gases emanating from volcanoes and mineral springs, to the products of combustion, and to respiration of animals, plants, and soil and marine microörganisms. On the debit side there are the assimilation of carbon by plants and the absorption of carbon dioxide in the weathering of certain rocks. Table V is a summary of the data compiled from different sources.

TABLE V. QUANTITATIVE DATA ON THE CARBON CYCLE (CARBON RESOURCES OF THE EARTH)

[In billions (10^9) of metric tons]

		CITATION
Weight of:		
Earth's crust (16 km. deep)	20,000,000,000	(42)
Hydrosphere	1,280,000,000	(10)
Atmosphere	5,373,000	(10)
Insoluble carbon in:		
Igneous rocks	52,139,000	(10)
Limestone	5,455,000	(10)
Shale	5,717,000	(10)
Sandstone	2,050,000	(10)
Available carbon in:		
Hydrosphere	16,400	(20, 21)
Atmosphere	600	(10)
Anthracite (85% C)[a]	422	(23)[b]
Bituminous coals (70% C)[a]	2,732	(23)[b]
Brown coals (50% C)[a]	1,499	(23)[b]
Peat	1,123	[c]
Petroleum	7	(13)
Soils (humus) (30 cm. deep)	400	[d]
All living matter	700	(42)
Total insol. carbon	65,361,000	
Total available carbon	23,883	
Total carbon	65,384,883	

[a] Calculated from Clarke (10, pp. 764–72).
[b] From this reference are the following figures (in millions of metric tons) anthracite, 496,864; bituminous coal, 3,902,944; brown coal, 2,997,763.
[c] See Table VI.
[d] Based on an assumption of 100,000,000 sq. km. of fertile land and 13 kg. of carbon per cubic meter of top soil.

The earth's resources of carbon include the materials in the solid crust, in the hydrosphere, and in the atmosphere. The relative importance of the three sources of carbon should be considered in the light of availability, as well as of quantity. It is obvious that the carbon in igneous rocks, limestone, shale, and sandstone is not available for supplying the immediate needs of vegetation. On the other hand, the carbon listed as "available" includes material immediately and potentially accessible.

The hydrosphere is the largest source of available carbon, followed in order by that contained in the coal deposits. The atmosphere contains about 600 billion metric tons of carbon equivalent, which is substantially less than that present in coal and peat deposits, and only 50 per cent greater in amount than that found in soils. There is scarcely any need to emphasize the fact that the carbon in coal, peat, petroleum, and soil organic matter was at one time or another withdrawn from the atmosphere by living organisms. This particular point may be further clarified by the data in Table VI representing an estimate of the peat deposits.

The peat deposits of Europe, Asia, and North America, as indicated by incomplete figures, cover an area of more than 70 million hectares. Basing our estimates on the figures given by Früh and Schroeter (*14*), we find that this area, to an average depth of 8 meters, contains 5,616,800 million cubic meters. Assuming further that the average dry weight of one cubic meter of peat is 400 kg. (*9*) and an average carbon content of 50 per cent (*33*), the world's peat deposits will be equivalent to at least 1123 billion metric tons.

TABLE VI. KNOWN PEAT DEPOSITS OF THE NORTHERN HEMISPHERE

	PEAT	CITATION
	Hectares	
Europe:		
Russia	19,000,000	(*37*)
Norway	1,600,000	(*19, 41*)
Sweden	5,000,000	(*17*, p. 61; *38*)
Finland	10,000,000	(*29*)
Denmark	100,000	(*17*, p. 61)
Holland	176,000	(*17*, p. 58)
Switzerland	6,000	(*14*)
Austria	34,000	(*35, 44*)
Germany	2,294,000	(*30*)
All others	1,000,000	...
Asia:		
Turkestan	7,000,000	
Transcaucasia	1,400,000	(*37*)
Siberia	10,000,000	
America:		
United States	3,600,000	(*12*)
Canada	9,000,000	(*31*)
Total	70,210,000	

Some of the quantitative relations in the carbon cycle are indicated in Table VII.

The atmosphere receives additions of carbon dioxide through the burning of coal, petroleum, wood, peat, crop residues, etc. Further additions are to be credited to respiration by the human and animal population. The most important credit item is that of the carbon dioxide evolved from soils. Bacteria, fungi, protozoa, and insects are responsible for the

TABLE VII. QUANTITATIVE DATA ON THE CARBON CYCLE
(CARBON IN CIRCULATION)

(In millions of metric tons)

	CARBON	CITATION
Additions:		
Combustion	900	(18)[a]
Respiration:		
Man 163		
Animals 1022		
Total	1,185	[b]
Soil respiration	19,000	(22)
Volcanoes	?	..
Losses:		
Assimilated by plants	16,300	(39)
Rock weathering	2,500	
Insol. carbonates	250	(42)
Ocean life	5,000	

[a] Clarke in 1919 estimated the annual coal consumption at a billion tons, while A. Krogh in 1904 estimated it at 700 million metric tons. The average carbon content of coal is, according to Lundegardh (22), 75 per cent.

[b] Assuming 1600 million adults of average weight 70 kg. and 280 grams of carbon per adult per day (40), and 1000 million tons of animals and 2800 grams of carbon per ton per day (3).

evolution of a vast quantity of carbon dioxide from soil organic matter. Limitations of space will not permit a more detailed discussion of the subject. It may be noted, however, that no definite information is available on the total quantity of carbon dioxide and other gases containing carbon present in the emanations from volcanoes and mineral springs. Boussingault (7) estimated in 1844 that the volcano Cotopaxi exhaled 500,000 tons of carbon as carbon dioxide. Under the head of debits, we may note the assimilation annually of more than 16 billion tons of carbon by vegetation. There is the fixing of large amounts of carbon in the weathering of rocks and in the formation of insoluble carbonates. An equivalent of 5 billion tons of carbon is fixed annually through the life processes of sea organisms. Altogether, it appears that the carbon dioxide withdrawn from the atmosphere by plants is balanced by that added to the air by microörganic and other processes in the soil. By way of further clarification, we should note that the carbon given off by the soil comes partly from the destruction of organic matter and partly from root respiration. According to the investigations of J. W. Shive, the root respiration of a corn crop amounts to about 35 per cent of the quantity of carbon fixed as organic matter by this same corn crop. If this ratio is extended to the entire plant world, it would mean that about 7,000 of the 19,000 million tons of the carbon addition diffusing from soils can be credited to root respiration, leaving 12,000 million tons due to the destruction of organic matter. If we add the 7,000 million tons of carbon transpired by roots to the 16,300 million tons of the carbon assimilated by plants, we obtain a total of 23,300 million tons of carbon utilized annually by vegetation. The release through soil respiration is but 19,000 million tons.

Oxygen and Hydrogen

Hydrogen and oxygen travel as water from the sea and the earth to the air. The living human and animal population of our world has a total weight of at least 1500 million metric tons; it carries about 750 to 800 million metric tons of uncombined water. This would form a lake 30 square miles in area and 33 feet deep. There is even a greater amount of uncombined water in living plants. It would be represented by something like 500 billion metric tons (*39*).[1] As to combined oxygen and hydrogen, the living plant population contains 262 billion metric tons of the former and 36 billion metric tons of the latter. The corresponding figures for the animal world are 122 million and 61 million metric tons, respectively. These figures are aside from the oxygen and hydrogen present as uncombined water in plant residues in soils, and likewise aside from the oxygen and hydrogen chemically combined in these residues.

Nitrogen and Sulfur

The primary source of nitrogen for vegetation is the atmosphere. Combined nitrogen in the form of ammonia, nitrites, nitrates, and organic matter is brought down to the land in atmospheric precipitation. Numerous determinations have shown that the nitrogenous materials thus brought down represent an equivalent of about 5 pounds per acre per year. In the course of many centuries these additions, conserved by the vegetation of forest and grassland, would result in a vast accumulation of soil nitrogen. There are, how-

[1] Schroeder estimates the carbon dioxide bound in the total standing green vegetation of the earth as follows (in billions of metric tons): forest leaves, 60; forest wood, 1000; cultivated land vegetation, 28; prairie vegetation, 8; desert vegetation, 2. As total carbon bound in vegetation he gives 275 ± 50 billion metric tons. From these figures it was possible to calculate the total organic matter in plants at 1100 billion tons.

ever, compensating factors of loss which will be mentioned later. Another factor on the credit side is the fixation of nitrogen by associations of bacteria and plants, or by bacteria alone. Where native vegetation is not disturbed by human activities, the accumulation of nitrogen in the soil will continue until the maximum is reached for any given soil type and climate. Thereafter, the positive factors of atmospheric precipitation and nitrogen fixation will be balanced by the negative factors of erosion, leaching, burning, grazing, etc. The plowed depth of an acre of land of good quality contains about one ton of nitrogen; in the entire profile it may be equivalent to 4000 or 5000 pounds. Fertile mineral soils may contain two or three times the amount just noted. In peat soils the amount is greater still.

When land is made arable, the factors of nitrogen accumulation are often overshadowed by those of nitrogen dissipation. The supply of soil nitrogen is then gradually depleted. If the process of nitrogen mining is carried far enough, the crop yields fall to a low level and the well-being of the animal and human populations on such land is more or less seriously threatened. A condition of this sort has been vividly described by Prothero (*36*) for the period 1300–1485:

> There was little to mitigate, either for men or beasts, the horrors of winter scarcity. Nothing is more characteristic of the infancy of farming than the violence of its alternations. On land which was inadequately manured, and on which neither field turnips nor clovers were known till centuries later, there could be no middle course between the exhaustion of continuous cropping and the rest cure of barrenness.

The extent of nitrogen losses from cultivated soils may be illustrated by the nitrogen studies of this station. In a series of cylinder experiments begun in 1898, continuous cropping, for a period of 35 years, resulted in a net loss equivalent to nearly one ton of nitrogen per acre, or about half of the total present at the beginning of the experiment. In a similar study conducted in field plots, the initial nitrogen content of the soil was 0.11 per cent. At the end of 25 years one of the plots showed a soil nitrogen content of only 0.065, or a loss of 900 pounds per acre from land initially poor in nitrogen.

Atmospheric precipitation adds substantial amounts of

TABLE VIII. ANNUAL ADDITIONS AND LOSSES OF PLANT NUTRIENTS IN SOILS OF THE UNITED STATES[a]

(In thousands of tons)

	N	P	K	Ca	Mg	S	Organic Matter
			LOSSES				
Crops (harvested areas)	4,600	700	3,200	1,000	500	500	92,000
Grazing (pastures)	3,000	500	3,700	1,000	500	400	60,000
Leaching (harvested areas)	4,000	....	6,600	26,600	6,000	7,400	80,000
Erosion (harvested areas)	2,500	900	15,000	13,000	6,000	800	50,000
Leaching (pastures)	1,000	....	1,700	7,000	1,600	1,900	20,000
Erosion (pastures)	1,000	400	6,000	5,000	2,200	300	20,000
Destruction of organic matter				Not determined			
Additional losses							
Total	16,100	2,500	36,200	53,600	16,800	11,300	322,000
			ADDITIONS				
Resources of 6⅔-inch top soil	1,700,000	800,000	12,000,000	12,000,000	5,500,000	450,000	34,000,000
Fertilizers and liming materials	300	300	300	2,100	Included with Ca	700	Small amounts
Manures	2,600	800	2,000	1,200	600	300	100,000
Rainfall (1318 million acres)	2,750	206	2,750	4,125	1,030	7,050	Not detd.
Irrigation (20 million acres)	30	....	500	3,000	1,000	2,200	Not detd.
N fixation by legumes	900	....	....	....	....	....	....
N fixation in other ways	6,800	....	....	....	....	....	....
Total additions	13,380	1,306	5,550	10,425	2,630	10,250	100,000
Net losses	2,720	1,194	30,650	43,175	14,170	1,050	222,000

[a] Area considered: harvested crops, 365 million acres; pastures and woodland grazing areas, 1000 million acres.

sulfur to the land. These additions vary quantitatively with the amount and distribution of rainfall. A careful survey of the available data indicates that the average of the 403 million crop acres and 914 million acres of pasture and range lands receives annually an equivalent of 10 to 11 pounds of sulfur. For the entire area this would represent a total of more than 7 million tons of sulfur. There are other additions to be credited to animal manures and chemical fertilizers. On the other hand, there are losses chargeable against the removal of sulfur in crops, animals, and animal products. Other losses are due to erosion, leaching, combustion, and certain bacterial activities. Table VIII summarizes the losses and gains of plant-nutrients in the United States.

While the data in Table VIII tell their own story, we should not fail to stress duly the significance of the figures, as they relate to the present and future ability of our agricultural areas to support plant and animal populations.

CHEMICAL NATURE OF SOILS

Soils are the culture media of terrestrial plants. The lithosphere, hydrosphere, and atmosphere contribute to their formation, as do also plants and other living organisms. In the soil, plants find their ash ingredients and a supply of water which is in itself a source of building materials, a solvent of soil constituents, a vehicle for the translocation of plant nutrients, and a temperature regulator.

The types and varieties of soil are numerous. They may be mineral or organic in character. They may be made up largely of silica or silicates. They may be poor or rich in colloids. They may be acid, neutral, or alkaline. They may contain a large or small proportion of compounds of calcium, magnesium, iron, aluminum, manganese in addition to potassium, sodium, and phosphorus. Their importance is much greater than their quantitative contribution to the volume of plant and animal products. Table IX includes analyses of soils differing widely as to origin and composition.

We may classify soils as mineral or organic in character. The latter are represented by reclaimed marsh areas, often very productive, but quantitatively of subordinate importance. They may contain 60 to 97 per cent of organic matter, usually more than 2 per cent of nitrogen, about 0.1 per cent of phosphorus, often less than 0.1 per cent of potassium, and a relatively high or low proportion of calcium. The Everglades of Florida and the high and low moor soils of the Netherlands and Germany are outstanding examples of organic soils.

The mineral soils include all but a small fraction of the cultivated soils of the several continents. They are the resultant of rock disintegration with time, climate, and living organisms as major factors of their genesis and development. The silica content of mineral soils usually represents a range of 60 to 95 per cent, with alumina and iron oxide as two other major ingredients. For the purpose of increasing crop yields, the supply of certain plant nutrients in the soil is supplemented by additions of materials furnishing nitrogen, phosphorus, potassium, and calcium. Now and then production may be increased by the use of materials containing sulfur, magnesium, and manganese. In the case of some soils, copper, iodine, boron, and two or three other ingredients have produced marked increase in yield.

Good mineral soils seldom contain less than 0.5 per cent of potassium, 0.04 per cent of phosphorus, and 0.03 to 0.04 per cent of sulfur. In the case of calcium and magnesium there is a twofold function—namely, supplying building material and combining with mineral and organic acids formed in the soil. Deficiencies of certain plant nutrients are reflected in the stunting of plants and may influence the growth and functioning of animals. The literature of soil and plant science is rich in data on the so-called deficiency diseases of plants and animals. McHargue and his associates have studied the significance of copper, manganese, zinc, iodine, and boron in the nutrition of microörganisms, plants, and animals. They found 14.04 parts of iodine per billion in the water supply of Lexington, Ky. (*26*). They also report (*27*) the determination of iodine in 439 samples of soil drawn from the important geological areas of Kentucky. Their analyses show a range of about 1 to 17 parts of iodine per million of soil. They offer evidence that boron is essential for the growth of lettuce (*23*); that the leaves of twenty-three species of deciduous forest trees which they examined contained copper, manganese, and zinc (*25*); and that copper, manganese, iron, "as well as the major constituents calcium,

TABLE IX. CHEMICAL COMPOSITION OF SOME REPRESENTATIVE SOIL TYPES[a]

(In per cent of dry weight)

Soil	Source	N	P	K	Ca	Mg	S	Si	Al	Fe	Mn	Ti	Na
Gray	Turkestan	0.08	0.09	2.13	5.17	1.86	0.21	27.96	5.92	3.63	...	...	...
Gray	Idaho	...	0.07	1.43	1.45	1.88	0.05	33.84	6.43	2.74	0.05	0.38	1.31
Chestnut brown	(Glinka)	...	0.07	...	1.75	1.29	...	29.33	7.94	3.56	...	...	1.62
Brown	Russia	0.21	0.03	...	0.06	0.83	0.19	34.31	5.43	2.64	...	...	...
Chernozem	Canada	0.55	0.08	1.88	1.12	0.48	...	32.62	5.25	1.80	...	...	...
Chernozem	Russia	...	0.10	1.40	0.93	...	0.004	20.72	8.35	3.16	0.05	...	0.53
Chernozem	N. Dak.	0.59	0.09	1.10	4.02	1.42	0.14	29.15	3.91	1.95	...	...	...
Prairie	Iowa	...	0.08	1.12	0.60	0.34	0.04	36.10	4.72	2.02	0.04	0.32	0.85
Brown forest	(Miami silt loam)	0.33	0.08	1.63	0.51	0.37	0.05	33.55	4.79	2.03	0.10	0.34	0.79
Gray forest	......	0.05	0.02	1.26	0.32	...	...	38.48	4.63	1.38	...	...	...
Gray forest	......	0.23	0.03	1.80	0.65	0.70	...	38.17	6.84	2.69	...	...	...
Podzol	......	0.21	0.04	2.22	0.59	0.33	0.01	32.19	4.72	2.75	0.04	0.20	0.79
Red and yellow earth	Georgia	...	0.01	0.13	0.43	0.08	...	43.48	6.44	0.77	0.04	0.20	0.16
Laterite	Cuba	...	0.01	0.05	0.09	0.20	...	1.53	9.77	44.09	0.33	0.48	0.36
Sphagnum peat	Maine	0.78	0.04	0.25	0.34	...	0.08	...	0.35	0.21	...	...	...
Low-lime peat	Minnesota	2.22	0.08	0.06	0.28	0.14	...	...	...	...	0.02	0.02	0.08

[a] Compiled by J. S. Joffe.

magnesium, phosphorus, and potassium, are factors in protecting pigeons from polyneuritis" (*24*). In eight different grasses grown in Idaho, Bolin (*6*) reported that "the average manganese content (dry basis) ranged from 207.5 mg. per kg. for orchard grass to 78.1 mg. per kg. for Kentucky blue grass. Alfalfa, with an average of 46.6 mg. per kg., was lower in manganese than any of the grasses. The eight grasses varied markedly in their capacity to extract manganese from the soil."

The meaning of these and of similar data becomes clearer in the light of the investigation conducted by Orr and his associates at the Rowett Research Institute, Aberdeen (*32*). They point out that "pasture is the natural food of the species of herbivora which have been domesticated. It is therefore the natural raw material of the most important animal products—viz., milk, meat, wool, and hides. Indeed, the greater part of the world supply of these primary necessities is drawn from animals whose food consists entirely of pasture, either grazed direct or preserved as hay."

Soil deficiencies are reflected in abnormal composition of pasture grasses. To quote Orr again:

In cases of extreme poverty, lack of one or other of the minerals may be the cause of disease in the grazing animal. The pathological conditions due to these deficiencies have received various names in different parts of the world, but the symptoms and lesions fall into definite groups according to the nature of the deficiency. Thus "styfsiekte" in South Africa, "coastal disease" in Australia, "creeps" in Texas, and "Waihi disease" in New Zealand, which are all due to deficiency of phosphorus, have essentially the same symptoms. In the same way osteoporosis due to lack of calcium, goiter due to lack of iodine, or anaemia due to lack of iron, present similar symptoms in whatever part of the world they occur. A future nomenclature will connect the disease with the cause.

In describing the depletion of some soil areas in Scotland, Orr says:

The factors which affect the mineral content of the herbage have a direct bearing on practical pasture problems. The composition of the soil can be improved by the application of fertilizers. It can also be changed in the opposite direction by depletion. This process of depletion and the resulting deterioration which shows itself in decreased rate of growth and production, and in extreme cases by the appearance of disease, is proceeding on all pastures from which milk, carcasses, or other animal products are taken off without a corresponding replacement being made. Accompanying the visible movement of milk and beef, there is a slow invisible flow of fertility. Every cargo of beef or milk products, every shipload of bones, leaves the exporting country so much the poorer. In many of the grazing areas of the world this depletion has become a serious economic problem. In Scotland, for example, generation after generation of sheep have been taken off the hills with little compensatory return...... This impoverishment of the poorer lands which has been going on for the last 200 years must be taken account of in considering the political question of depopulation. Before the population can be increased, it will be necessary to get back into the land the lime and phosphorus which have been taken off it. This is one of the most important political questions in Scotland.

The social, economic, and political meaning of soil depletion was understood in ancient times. The history of the Mediterranean basin bears witness to this fact. When the soils of Italy began to fail, the Roman legions reached out farther and farther for soil fertility on which they could levy tribute. Finally, circuses without bread did not save Rome. But Roman farmers, under the pressure of necessity, had discovered various expedients for offsetting soil deterioration, at least in part. Lupines and vetches as nitrogen fixers, calcareous marl as a source of calcium and, in some degree, of potassium and phosphorus, and various refuse vegetable and animal products as sources of nitrogen and of other plant nutrients were employed successfully toward maintaining soil fertility. The particular meaning of bones as a source of phosphorus for depleted soils was not fully understood until the end of the eighteenth century. In the last quarter of the eighteenth and in the first half of the nineteenth century the use of bones reached notable proportions. This led Baron Liebig to protest in the following words (1):

England is robbing all other countries of the condition of their fertility. Already, in her eagerness for bones, she has turned up the battle fields of Leipzig, of Waterloo, and of the Crimea; already from the catacombs of Sicily she has carried away the skeletons of many successive generations. Annually she removes from the shores of other countries to her own the manurial equivalent of three millions and a half of men, whom she takes from us the means of supporting, and squanders down her sewers to the sea. Like a vampire, she hangs upon the neck of Europe—nay the entire world—and sucks the heart blood from nations without a thought of justice towards them, without a shadow of lasting advantage to herself.

Those were strong words, spoken at a time when the enormous deposits of mineral phosphates had not yet been discovered. These, too, are the transformed bones of other geologic days. It is still true that nations, like dogs, will fight over the bones of yesterday and of today. In fighting for a place in the sun they reach out for more bread, a part of whose ingredients came from bones and stones.

Let us remember that we, too, have imported bones from Europe, only they were the bones and flesh of the living. Many millions of them have come since the founding of Jamestown. With their bodies they brought something of the composite human spirit whose vibrations we cannot escape. Our soils are still young, but part of them, like the old gray mare, "aint what she used to be." Our sewers carry away enormous quantities of plant nutrients. Only a minute portion of this soil fertility is salvaged. The extent of this removal is indicated by the following figures which show the gross annual loss of plant nutrients in sewage in the United States (in thousands of tons):[a]

Nitrogen	671.9	Potassium	117.6
Carbon	5513.0	Calcium	23.3
Phosphorus	69.6	Magnesium	21.8
Sulfur	57.9		

[a] Compiled with the assistance of estimates from the Department of Water Supplies and Sewage Disposal.

The loss of nitrogen is more than double that supplied by commercial fertilizers. Mention may also be made here of the nitrogen and ash ingredients which we have been exporting in successive decades since the census of 1790. While complete information on the subject is not available, we know that in an estimated 90 per cent of our agricultural exports we have alienated from the soils of the United States something like 12 million tons of nitrogen, 2 million of phosphorus, 3 million of potassium, 1 million of calcium, more than 0.75 million of magnesium, and 0.5 million of sulfur. He who has a flair for figures may attempt to estimate how many of the king's horses and men were built out of raw materials supplied by our soils.

In measuring the ability of land to support man and beast, we speak of "carrying capacity." There are vast differences in the carrying capacity of soil regions both in respect to vegetation and to animals and man. Rainfall is one of the major determining factors. This may be illustrated by the figures given in Table X.

TABLE X. AVERAGE ANNUAL PRECIPITATION IN THE UNITED STATES (43)

PRECIPITATION Inches	LAND AREA Acres	% OF TOTAL LAND AREA OF U. S.
Under 10	153,634,432	8.1
10–20	588,775,719	30.9
20–30	314,258,301	16.5
30–40	320,089,545	16.8
40–50	324,846,189	17.1
50–60	160,366,829	8.4
60–80	28,898,105	1.5
80–100	9,430,528	0.5
Over 100	2,915,712	0.2

More than 8 per cent of the land area of the country receives, on the average, less than 10 inches of annual precipitation. At the other extreme is an area of nearly 3 million acres receiving more than 100 inches. Fortunately for us, about 60 per cent of our entire land surface has an annual rainfall of 20 to 80 inches. We should be reminded again that one inch of rainfall is equivalent to about 100 tons of water per acre, and that water is used as a source of oxygen and hydrogen for the manufacture of carbohydrates, fats, proteins, and other plant products. This is aside from the part played by uncombined water in the functioning of plants and animals.

By "carrying capacity" is meant the ability of the land to support plant and animal populations. In a narrower sense carrying capacity refers to the capacity of pasture and range lands to furnish food to domestic animals during the grazing season. Climate, season, soil type, and soil management are major factors in determining carrying capacity. According to Piper (34) much of the western range lands have a carrying capacity of one steer to 100 acres for the entire year. The better range lands have a carrying capacity of one steer to 25 acres. The best blue grass pasture will carry one steer to 2.5 acres for the grazing season of 5 to 6 months. The best Bermuda grass and lespedeza pastures will carry 2 steers to one acre for the grazing season. What about the carrying capacity of the land in earlier geologic times? If, at the present time, a 1000-pound steer requires 25 acres of good range land for his support, a 30-ton giant dinosaur would require, on the same basis, something like 1500 acres, a good deal of territory to cover in search of food.

The ability of the land to support human population is dependent, aside from the factors already mentioned, on dietaries and standards of living. It is estimated by Baker (4) that the per capita use of land in the United States is equivalent to about 5 acres of improved land. This may be compared with less than 1 acre in some of the oriental countries. In this connection, Alsberg (2) stated that "despite the inefficiency of the green plant, the quantities of energy stored up by agriculture in such a country as the United States as food for man and fodder for domesticated animals is enormous—at least the equivalent of 10 to 20 per cent of our entire annual coal production, anthracite and bituminous combined." All told, then, the soil and the air must furnish the raw materials of the plant and animal organisms. The supply of these and the conditions under which they are assimilated will of necessity reflect the prevailing social and economic conditions.

Finally, something should be said here about the salvaging and re-use of organic debris. Living protoplasm has constructed, in the course of unnumbered centuries, animal and plant products which, if they had been allowed to accumulate, would have reached the proportions of great mountain ranges. But the supply of atmospheric carbon would have been exhausted within a few years. Lundegardh claims that this would come to pass within a period of 35 years. Fortunately bacteria, and other microörganisms, as well as higher plants and animals, provide the mechanism by virtue of which carbon, nitrogen, and sulfur circulate between soil and air. Organic remains are attacked by invisible wrecking crews, plant and animal tissues crumble and vanish, the air comes into its own again, and, as to the ash ingredients, dust returns to dust. From the beginning of biological time to the very end of it, organic scrap has been, and will be, used over and over again. We who are here now are clad in the garment of life that has been worn in many lands, in many forms, in days without end. We come and we vanish, but we cannot say that the earth will know us no more. The garment we may cast aside, the rest of us cannot be destroyed. Having lived, ever so briefly, we leave behind us memories, faint or vivid, which join into one the past, the present, and the future.

LITERATURE CITED

(1) Aikman, C. M., "Manures and Manuring," p. 360, Edinburgh, Wm. Blackwood & Sons, 1894.
(2) Alsberg, C. L., IND. ENG. CHEM., 16, 524 (1924).
(3) Armsby, H. P., and Friess, J. A., U. S. Dept. Agr., Bur. Animal Ind., *Bull.* 128 (1911).
(4) Baker, O. E., *Geograph. Rev.*, 13, 1 (Jan., 1923).
(5) Bear, F. E., "Soil Management," New York, John Wiley & Sons, 1924; Gortner, R. A., "Outlines of Biochemistry," New York, John Wiley & Sons, 1929; Honcamp, F., Handbuch der Pflanzenernährung und Düngerlehre, Vol. 1, Berlin, Julius Springer, 1931; Lawes and Gilbert tables on composition of whole animal carcasses, quoted by Henry and Morrison (*16*) and others; Lotka, A. E., "Elements of Physical Biology," Baltimore, Williams & Wilkins, 1925; Mangold, Ernst, Handbuch der Ernährung u. s. w., Vols. 1–3, Berlin, Julius Springer, 1929.
(6) Bolin, D. W., *J. Agr. Research*, 48, 657–63 (1924).
(7) Boussingault, J., *Ann. chim. phys.*, [3] 10, 456–69 (1844).
(8) Carlyle, Thomas, "Sartor Resartus," Boston, Ginn and Co.
(9) Christensen, H. R., "Landøkonomisk Jordbundslaere," Copenhagen and Oslo, Gyldendahl, 1917.
(10) Clarke, F. W., Geol. Survey, *Bull.* 770 (1924).
(11) *Ibid.*, pp. 22, 36, 45.
(12) Davis, C. A., *Eng. Mag.*, No. 37, 80–9 (1909).
(13) Dewhurst, T., *J. Inst. Petroleum Tech.*, 20 (126), 280 (1934).
(14) Früh, J., and Schroeter, C., "Die Moore der Schweiss, mit Beruchsichtigung der gesamten Moorfrage. Beitrage zur geologischen Karte der Schweiss," Geotechnische Serie No. 1, 1904.
(15) Hart and Tottingham, "General Agricultural Chemistry," p. 103, Madison, Wis., 1923.
(16) Henry and Morrison, "Feeds and Feeding," 19th ed., p. 710, Madison, Wis., Henry and Morrison Co., 1923.
(17) Hoering, Paul, "Moornutzung und Torfverwerkung," Berlin, Julius Springer, 1915 (reprint, 1921).
(18) Holbrook, E. A., *Ann. Acad. Political Social Sci.*, 111, 203 (1924).
(19) Holmboe, Jens, in Englers Bot. Jahrbuch, Vol. 34, Heft 2, p. 207 (1904).
(20) Krogh, A., *Compt., rend.*, 139, 896 (1904).
(21) Linck, G., Handwörterbuch der Naturwissenschaften, Vol. 5, p. 1049, Jena, G. Fischer, 1914.
(22) Lundegardh, Henrik, "Der Kreislauf der Kohlensäure," Jena, G. Fischer, 1924.
(23) McHargue, J. S., and Calfee, R. K., *Plant Physiol.*, 8, 305–13 (1933).
(24) McHargue, J. S., and Roy, W. R., *Am. J. Physiol.*, 99, 221–6

(1931); McHargue, J. S., and Calfee, R. K., *Plant Physiol.*, 4, 697–703 (1932); *Botan. Gaz.*, 91, 183–93 (1931).
(25) McHargue, J. S., and Roy, W. R., *Botan. Gaz.*, 94, 381–93 (1932).
(26) McHargue, J. S., and Young, D. W., *Am. Water Works Assoc.*, 25, 380–2 (1933).
(27) McHargue, J. S., and Young, D. W., *Soil Sci.*, 35, 425–35 (1933).
(28) McInnes, Wm., and others, "Coal Resources of the World," Vol. 1, XII. Internat. Geol. Congress, Canada, 1913, Toronto, Morang & Co., 1913.
(29) Malm, G. A., "Eine Karte über die Moore des südlichen Hälfte von Finland," Helsingfors, 1912.
(30) Meitzen, A., "Der Boden und die landwirtschaflichen Verhältnisse der Preussischen Staates," Berlin, 1868.
(31) Nyström, E., "Peat and Lignite," Ottawa, 1908.
(32) Orr. J. B., and others, "Minerals in Pastures and Their Relation to Animal Nutrition," pp. 138 ff., London, Lewis & Co., 1929.
(33) Pia, J., "Pflanzen als Gesteinsbildner," Berlin, Gebrüder Borntraeger, 1926.
(34) Piper, C. V., "Forage Crops and Their Culture," pp. 108–9, New York, Macmillan Co., 1914.
(35) Pokorny, A., *Sitzber. Kaiserl. Akad. Wiss.*, 43, 57; *Verhandl. Kaiserl. Königl. zool-bot. Ges. Wien*, 8 (1858), 9 (1859).
(36) Prothero, R. E., "English Farming Past and Present," 2nd ed., p. 33, London, Longmans, Green & Co., 1917.
(37) Schreiber, H., in *Oesterreichische Moorzeitschrift Staab.*, 1905.
(38) Schreiber, H., *Ibid.*, 1914.
(39) Schroeder, H., *Naturwissenschaften*, 7, 8, 23 (1919).
(40) Sherman, H. C., "Chemistry of Food and Nutrition," 4th ed., New York, Macmillan Co., 1932.
(41) Thaulow, *Meddelelser fra det Norske Myrselskab*, Heft 1, p. 19, Oslo, 1905.
(42) Vernadsky, W. J. (tr. into German by E. Kordes), "Geochemie," pp. 140–232, Leipzig, Akademische Verlagsgesellschaft, 1930.
(43) Yearbook of Agriculture, 1921, p. 418.
(44) Zailer, V., in *Jahrbuch der Moorkunde*, p. 51 (1913); *Erhnährung der Pflanze*, 34–8 (1912).

COLUMBIA UNIVERSITY PRESS
COLUMBIA UNIVERSITY
NEW YORK

FOREIGN AGENT
OXFORD UNIVERSITY PRESS
HUMPHREY MILFORD
AMEN HOUSE, E. C.
LONDON

Bei Fragen zur Produktsicherheit wenden Sie sich bitte an:
If you have any questions regarding product safety,
please contact:

Walter de Gruyter GmbH
Genthiner Straße 13
10785 Berlin
productsafety@degruyterbrill.com